HOW TO USE A SCIENTIFIC CALCULATOR

Unlocking Its Full Potential for Calculations beyond Basic Math

James Roland

All rights reserved. No part of this publication may be reproduced in any form or by any means, including photocopying, recording, or any other electronic or mechanical methods without the prior written permission of the publisher except in the case of brief quotations embodied in reviews and certain other non-commercial uses permitted by copyrights law.

Copyright © James Roland, 2024.

TABLE OF CONTENTS

Chapter 1 9
Introduction to Scientific Calculators 9
   1.1 What is a Scientific Calculator? 9
   1.2 Types of Scientific Calculators 10
   1.3 Basic Calculator Anatomy: Keys and Displays 11
   1.4 Calculator Modes: Degrees, Radians, Gradians
   13
   1.5 Common Mistakes to Avoid: Order of
   Operations 14
   1.6 Calculator Maintenance and Care Tips 15
   1.7 Exploring Different Calculator Brands 16
Chapter 2 18
Essential Arithmetic Operations 18
   2.1 Addition, Subtraction, Multiplication, and
   Division 18
   2.2 Powers and Roots: Squaring, Cubing, Square
   Roots, Cube Roots 20
   2.3 Parentheses and Brackets: Grouping Operations
   22
   2.4 Fractions and Decimals: Converting and
   Calculating 23
   2.5 Percentages: Calculations and Conversions 24
   2.6 Scientific Notation: Understanding and Using 25

2.7 Memory Functions: Storing and Recalling Values 26

Chapter 3 29

Trigonometric Functions 29

   3.1 Sine, Cosine, and Tangent: Definitions and Applications 29

   3.2 Inverse Trigonometric Functions: Arcsine, Arccosine, Arctangent 30

   3.3 Degrees, Radians, and Gradians: Converting Between Units 31

   3.4 Solving Triangles: Right-Angle and Oblique Triangles 32

   3.5 Trigonometric Identities: Understanding and Applying 33

   3.6 Graphing Trigonometric Functions 34

   3.7 Practical Applications of Trigonometry: Physics, Engineering 35

Chapter 4 37

Logarithmic and Exponential Functions 37

   4.1 Logarithms: Definitions and Properties 37

   4.2 Common Logarithms (Base 10) and Natural Logarithms (Base e) 39

   4.3 Antilogarithms (Inverse Logarithms) 39

   4.4 Solving Logarithmic and Exponential Equations 40

    4.5 Applications of Logarithms: pH, Richter Scale, Decibel Scale..................41

    4.6 Graphing Logarithmic and Exponential Functions 43

    4.7 Exponential Growth and Decay: Real-World Applications 44

Chapter 5 46

Statistical Functions 46

    5.1 Mean, Median, and Mode: Measures of Central Tendency 46

    5.2 Range and Standard Deviation: Measures of Dispersion 47

    5.3 Probability Calculations: Permutations and Combinations 48

    5.4 Normal Distribution: Understanding and Applying 49

    5.5 Hypothesis Testing: Introduction and Basic Concepts 50

    5.6 Regression Analysis: Linear and Multiple Regression 51

    5.7 Statistical Applications: Data Analysis and Interpretation 52

Chapter 6 54

Complex Numbers 54

    6.1 Imaginary Unit (i) and Complex Number Form 54

6.2 Arithmetic Operations with Complex Numbers 55

6.3 Conjugate of a Complex Number 57

6.4 Modulus and Argument of a Complex Number 57

6.5 Polar Form of Complex Numbers 58

6.6 De Moivre's Theorem: Powers and Roots of Complex Numbers 59

6.7 Applications of Complex Numbers: Electrical Engineering, Quantum Mechanics 59

Chapter 7 61

Calculus Functions 61

7.1 Derivatives: Instantaneous Rates of Change 61

7.2 Basic Differentiation Rules: Power Rule, Product Rule, Quotient Rule 62

7.3 Integrals: Area Under a Curve 63

7.4 Basic Integration Rules: Power Rule, Substitution Rule 64

7.5 Numerical Integration: Approximating Definite Integrals 64

7.6 Differential Equations: Introduction and Basic Concepts 65

7.7 Applications of Calculus: Physics, Engineering, Economics 66

Chapter 8 68

Advanced Functions and Features 68

8.1 Hyperbolic Functions: Sinh, Cosh, Tanh 68

8.2 Matrix Operations: Addition, Subtraction, Multiplication, Determinants 69

8.3 Vector Calculations: Dot Product, Cross Product 70

8.4 Equation Solvers: Numerical and Symbolic Solutions 71

8.5 Programming Functions: Creating Custom Programs 72

8.6 Graphing Functions: 2D and 3D Graphs 72

8.7 Solver Functions: Financial Calculations, Unit Conversions 73

Chapter 9 75

Practical Applications in Various Fields 75

9.1 Science: Chemistry, Physics, Biology 75

9.2 Engineering: Electrical, Mechanical, Civil 77

9.3 Finance and Economics: Interest Calculations, Statistics 78

9.4 Computer Science: Algorithms, Data Analysis 79

9.5 Mathematics: Algebra, Geometry, Trigonometry 79

9.6 Surveying and Navigation: Coordinates, Distance Calculations 80

9.7 Everyday Life: Budgeting, Conversions, Time Calculations 81

Chapter 10 82

Troubleshooting and Tips 82

   10.1 Common Calculator Errors: Syntax, Calculation, Overflow 82

   10.2 Resetting Your Calculator 83

   10.3 Finding Manuals and Online Resources 84

   10.4 Upgrading Your Calculator: Firmware and Software 85

   10.5 Choosing the Right Calculator for Your Needs 85

   10.6 Utilizing Online Calculators and Apps 86

   10.7 Tips for Effective Calculator Use: Organization, Practice 87

Chapter 1

Introduction to Scientific Calculators

Welcome to the fascinating world of scientific calculators! While they may seem intimidating at first, these powerful tools are essential companions for students, scientists, engineers, and anyone who delves into mathematics beyond basic arithmetic. In this chapter, we'll demystify scientific calculators, exploring their definition, various types, essential components, modes, common usage errors, and maintenance tips. We'll also take a peek into the diverse landscape of calculator brands, helping you find the perfect fit for your needs.

1.1 What is a Scientific Calculator?

A scientific calculator is a specialized electronic calculator designed to perform a wide range of mathematical functions beyond basic arithmetic operations like addition, subtraction, multiplication, and division. Unlike standard calculators, which are often limited to these fundamental operations, scientific calculators can handle complex calculations involving exponents, logarithms, trigonometric functions (sine, cosine, tangent), and statistical functions.

The power of scientific calculators lies in their ability to simplify complex calculations, saving users significant time and effort. Whether you're solving equations, analyzing data, or working with advanced mathematical concepts, a scientific calculator is an invaluable tool that streamlines the process and ensures accuracy.

1.2 Types of Scientific Calculators

Scientific calculators come in various types, each catering to specific needs and levels of expertise. Let's explore some of the most common categories:

Basic Scientific Calculators: These entry-level calculators are suitable for students and individuals who require basic scientific functions. They typically include trigonometric functions, logarithms, exponents, and basic statistical operations.

Graphing Calculators: A step up from basic models, graphing calculators can plot functions, display graphs, and solve equations graphically. They are widely used in advanced mathematics, physics, engineering, and other scientific fields.

Programmable Calculators: These calculators allow users to create and store programs, enabling them to automate repetitive calculations or perform complex tasks. Programmable calculators are often used in engineering, scientific research, and other specialized fields.

Financial Calculators: Designed for financial professionals, these calculators include functions for calculating interest rates, amortization schedules, loan payments, and other financial metrics.

Specialty Calculators: Some calculators are tailored to specific fields, such as statistics calculators, chemistry calculators, and engineering calculators. These specialized calculators often include functions and features unique to their respective disciplines.

The choice of calculator type depends on your specific needs and budget. Consider the level of complexity of the calculations you'll be performing, the features you require, and your financial constraints before making a decision.

1.3 Basic Calculator Anatomy: Keys and Displays

Understanding the basic anatomy of a scientific calculator is crucial for efficient usage. Let's familiarize ourselves with the key components:

Display: The display is the screen where numbers, symbols, and results are shown. It can be a liquid crystal display (LCD) or a light-emitting diode (LED) display. Some calculators have multi-line displays, allowing them to show entire equations or multiple results simultaneously.

Keypad: The keypad contains the various buttons used for inputting numbers, operators, functions, and commands. The arrangement of keys can vary depending on the model and brand of the calculator.

Function Keys: These are the keys that represent various mathematical functions, such as sine (sin), cosine (cos), tangent (tan), logarithm (log), exponential (exp), and square root ($\sqrt{}$). Some calculators have dedicated keys for these functions, while others require pressing a "shift" or "2nd" key before accessing them.

Operator Keys: These include the basic arithmetic operators (+, -, x, /) as well as other operators like parentheses, brackets, exponents, and factorials.

Navigation Keys: These keys are used for moving the cursor, editing input, and accessing various menus or options on the calculator.

Mode Key: The mode key allows you to switch between different calculation modes, such as degrees, radians, or gradians (for angle measurements) and floating-point or fixed-decimal display modes.

Familiarizing yourself with the layout and function of the keys on your specific calculator is the first step towards mastering its use.

1.4 Calculator Modes: Degrees, Radians, Gradians

Scientific calculators can operate in different modes, particularly when it comes to angle measurements. The three most common modes are:

Degrees (°): This is the most familiar mode for angle measurement, where a full circle is divided into 360 degrees.

Radians (rad): In radians, a full circle is equal to 2π radians. Radians are often used in calculus, physics, and engineering, as they simplify many trigonometric calculations.

Gradians (grad): Less commonly used, gradians divide a full circle into 400 gradians.

It's crucial to ensure that your calculator is in the correct mode for your calculations. If you're working with angles in degrees, ensure your calculator is in degree mode; if you're dealing with radians, switch to radian mode. Failing to set the correct mode can lead to incorrect results.

Most calculators have a dedicated "Mode" key that allows you to switch between different modes. Refer to your calculator's manual for specific instructions on how to change modes.

1.5 Common Mistakes to Avoid: Order of Operations

One of the most common errors when using a calculator is failing to follow the correct order of operations. This can lead to inaccurate results, even if you input the numbers and operators correctly.

The order of operations dictates the sequence in which mathematical operations should be performed. A helpful acronym for remembering this order is PEMDAS (or BODMAS):

- **P**arentheses (or **B**rackets)
- **E**xponents (or **O**rders)
- **M**ultiplication and **D**ivision (performed from left to right)
- **A**ddition and **S**ubtraction (performed from left to right)

By adhering to this order, you can ensure that your calculations are accurate and consistent with mathematical principles.

For example, consider the calculation $3 + 5 \times 2$. Following the order of operations, we first perform the multiplication ($5 \times 2 = 10$) and then the addition ($3 + 10 = 13$). However, if we mistakenly performed the addition first ($3 + 5 = 8$) and then the multiplication ($8 \times 2 = 16$), we would get an incorrect result.

When using a scientific calculator, it's important to be mindful of the order of operations and use parentheses or brackets to clarify the order in which calculations should be performed.

1.6 Calculator Maintenance and Care Tips

Proper maintenance and care can extend the lifespan of your scientific calculator and ensure its optimal performance. Here are some tips to keep in mind:

- **Keep it Clean:** Regularly wipe down your calculator with a soft, dry cloth to remove dust and debris. Avoid using harsh chemicals or abrasive materials that could damage the display or keys.
- **Protect the Display:** When not in use, cover your calculator with a protective case or sleeve to prevent scratches and damage to the display.
- **Store it Properly:** Store your calculator in a cool, dry place away from direct sunlight and extreme temperatures. Avoid storing it in humid environments, as moisture can damage the internal components.
- **Replace Batteries:** If your calculator uses batteries, replace them regularly to ensure uninterrupted operation. Follow the manufacturer's instructions for battery replacement.
- **Avoid Dropping:** While modern calculators are built to be durable, dropping them can still cause damage. Handle your calculator with care to avoid accidental falls.

- **Read the Manual:** Familiarize yourself with the manufacturer's instructions for operating and maintaining your calculator. This will help you avoid common errors and maximize its performance.

By following these simple tips, you can keep your scientific calculator in excellent condition for years to come.

1.7 Exploring Different Calculator Brands

The market offers a wide array of scientific calculator brands, each with its unique features, strengths, and target audience. Some of the most popular brands include:

- **Texas Instruments (TI):** TI is a well-established brand known for its diverse range of scientific calculators, from basic models for students to advanced graphing calculators used in college and professional settings.
- **Casio:** Casio is another leading manufacturer of scientific calculators, offering a variety of models with features such as high-resolution displays, intuitive interfaces, and robust functionality.
- **Hewlett-Packard (HP):** HP is renowned for its high-quality calculators, particularly its graphing calculators, which are favored by engineers, scientists, and mathematicians.

- **Sharp:** Sharp offers a range of scientific calculators with advanced features, including equation solvers, matrix operations, and complex number calculations.
- **Other Brands:** Several other brands, such as Canon, Citizen, and Victor, also manufacture scientific calculators. These brands often offer competitive features and affordable options for students and professionals alike.

Choosing the right calculator brand depends on your individual needs and preferences. Consider factors such as functionality, ease of use, price, and brand reputation when making your decision.

In conclusion, this comprehensive overview of scientific calculators should equip you with the knowledge needed to embark on your journey of mastering this invaluable tool. In the following chapters, we will delve deeper into the specific functions and applications of scientific calculators, unlocking their full potential for solving complex mathematical problems in various fields.

Chapter 2

Essential Arithmetic Operations

Welcome to the foundation of mathematical calculations – arithmetic operations! In this chapter, we'll delve into the fundamental operations of addition, subtraction, multiplication, and division, exploring how to perform them efficiently on your scientific calculator. We'll then extend our knowledge to include powers, roots, grouping operations, fractions, decimals, percentages, scientific notation, and the handy memory functions of your calculator. Mastering these essential arithmetic operations will empower you to tackle a wide range of mathematical problems confidently.

2.1 Addition, Subtraction, Multiplication, and Division

The four basic arithmetic operations form the bedrock of mathematical calculations. Let's briefly recap each operation and how to perform them on your scientific calculator:

Addition (+): The process of combining two or more numbers to find their total sum.

To add numbers on your calculator, simply press the "+" key between the numbers you want to add and then press the "=" key to obtain the result.

Example: 2 + 3 + 5 = 10

Subtraction (-): The process of finding the difference between two numbers.

To subtract numbers on your calculator, press the "-" key between the numbers you want to subtract and then press the "=" key.

Example: 8 - 3 = 5

Multiplication (x or): The process of repeated addition of a number.

To multiply numbers on your calculator, press the "x" or "*" key between the numbers you want to multiply and then press the "=" key.

Example: 4 x 6 = 24

Division (/): The process of splitting a number into equal parts or finding how many times one number is contained within another.

To divide numbers on your calculator, press the "/" key between the numbers you want to divide and then press the "=" key.

Example: 15 / 3 = 5

While these operations seem straightforward, it's crucial to remember the order of operations (PEMDAS/BODMAS) to ensure accurate results. If your calculation involves multiple operations, perform them in the correct order: parentheses/brackets first, followed by exponents, then multiplication and division (from left to right), and finally addition and subtraction (from left to right).

2.2 Powers and Roots: Squaring, Cubing, Square Roots, Cube Roots

Beyond basic arithmetic, scientific calculators offer functions for calculating powers and roots, allowing you to explore exponential relationships and solve various mathematical problems.

Powers: A power represents repeated multiplication of a number by itself. For example, 2 raised to the power of 3 (2^3) means 2 x 2 x 2 = 8.

To calculate powers on your calculator, use the "^" key or the "x^y" key.

Example: 5 ^ 2 = 25 (5 squared)

Squaring (x^2): Squaring a number means raising it to the power of 2.

Most scientific calculators have a dedicated "x^2" key for squaring a number.

Example: $7^2 = 49$ (7 squared)

Cubing (x^3): Cubing a number means raising it to the power of 3.

Some calculators have a dedicated "x^3" key for cubing a number. If not, you can use the "^" or "x^y" key to raise the number to the power of 3.

Example: $3^3 = 27$ (3 cubed)

Roots: A root is the inverse operation of raising a number to a power. The most common roots are square roots and cube roots.

Square Root ($\sqrt{\ }$): The square root of a number is the value that, when multiplied by itself, equals the original number.

Most scientific calculators have a dedicated "$\sqrt{\ }$" key for calculating square roots.

Example: $\sqrt{25} = 5$ (the square root of 25 is 5)

Cube Root ($\sqrt[3]{\ }$): The cube root of a number is the value that, when multiplied by itself three times, equals the original number.

Some calculators have a dedicated "$\sqrt[3]{\ }$" key for calculating cube roots. If not, you can use the "^(1/3)" function to find the cube root.

Example: $\sqrt[3]{27} = 3$ (the cube root of 27 is 3)

Understanding powers and roots is crucial for working with polynomials, solving equations, and exploring exponential relationships in various fields, including physics, engineering, and finance.

2.3 Parentheses and Brackets: Grouping Operations

Parentheses ( ) and brackets [ ] are used to group operations within a calculation, indicating the order in which they should be performed. This is particularly important when dealing with complex expressions involving multiple operations.

When using parentheses or brackets, the operations within the innermost set of parentheses or brackets are performed first, followed by the next outer set, and so on. This ensures that calculations are performed in the correct sequence, avoiding ambiguity and ensuring accurate results.

For example, consider the expression 2 x (3 + 4). Without parentheses, we might mistakenly multiply 2 by 3 first and then add 4, resulting in an incorrect answer of 10. However, with parentheses, we clearly indicate that we should first add 3 and 4 (resulting in 7) and then multiply the sum by 2, yielding the correct answer of 14.

Scientific calculators allow you to use parentheses and brackets to group operations, ensuring that calculations are performed in the desired order.

Always be mindful of the placement of parentheses and brackets to avoid errors in your calculations.

2.4 Fractions and Decimals: Converting and Calculating

Fractions and decimals are two ways of representing parts of a whole. Fractions express a quantity as a ratio of two numbers (e.g., 3/4), while decimals represent a quantity using a decimal point (e.g., 0.75).

Scientific calculators can handle both fractions and decimals, allowing you to perform calculations involving these representations effortlessly.

Converting Fractions to Decimals: To convert a fraction to a decimal, simply divide the numerator (top number) by the denominator (bottom number) using your calculator.

Example: 3 / 4 = 0.75

Converting Decimals to Fractions: To convert a decimal to a fraction, express the decimal as a fraction with a denominator that is a power of 10 (e.g., 10, 100, 1000). Then, simplify the fraction if possible.

Example: 0.75 = 75/100 = 3/4

Calculating with Fractions: Scientific calculators often have a dedicated fraction key (ab/c) for inputting and calculating with fractions.

Example: 1/2 + 2/3 = 7/6

Calculating with Decimals: Use the decimal point (.) to input and calculate with decimals.

Example: 0.5 + 0.25 = 0.75

Being able to work with both fractions and decimals expands your calculation capabilities and allows you to solve problems in various contexts, from everyday calculations to scientific and engineering applications.

2.5 Percentages: Calculations and Conversions

Percentages are a way of expressing a number as a fraction of 100. The symbol "%" is used to denote a percentage. For example, 50% means 50 out of 100.

Scientific calculators provide functions for calculating percentages and converting between percentages, decimals, and fractions.

Calculating Percentages: To calculate a percentage of a number, multiply the number by the percentage and then divide by 100.

Example: 20% of 80 = (20 x 80) / 100 = 16

Converting Percentages to Decimals: To convert a percentage to a decimal, divide the percentage by 100.

Example: 50% = 50 / 100 = 0.5

Converting Decimals to Percentages: To convert a decimal to a percentage, multiply the decimal by 100.

Example: 0.75 = 0.75 x 100 = 75%

Converting Percentages to Fractions: To convert a percentage to a fraction, express the percentage as a fraction with a denominator of 100 and simplify if possible.

Example: 25% = 25/100 = 1/4

Percentages are widely used in various fields, including finance, statistics, and everyday life. Mastering percentage calculations and conversions is essential for understanding discounts, interest rates, taxes, and other real-world applications.

2.6 Scientific Notation: Understanding and Using

Scientific notation is a convenient way of expressing very large or very small numbers. It involves writing a number as a product of a decimal number between 1 and 10 and a power of 10.

For example, the speed of light, approximately 299,792,458 meters per second, can be expressed in scientific notation as 2.99792458×10^8 m/s.

Scientific calculators can handle numbers in scientific notation, making it easier to work with large or small quantities.

To enter a number in scientific notation on your calculator, use the "EE" or "EXP" key.

Example: To enter 2.99792458×10^8, press the following keys: 2.99792458 EE 8.

To display results in scientific notation, set your calculator to scientific mode using the "Mode" key. Refer to your calculator's manual for specific instructions on how to switch modes.

Scientific notation is widely used in scientific disciplines, particularly physics, chemistry, and astronomy, where dealing with extremely large or small quantities is commonplace. Understanding scientific notation is crucial for interpreting scientific data, performing calculations, and communicating results effectively.

2.7 Memory Functions: Storing and Recalling Values

Most scientific calculators come equipped with memory functions that allow you to store and recall values during calculations. This can be immensely helpful when dealing with complex calculations or when you need to use the same value multiple times.

Common memory functions include:

- **M+ (Memory Plus):** Adds the current value on the display to the stored value in memory.

- **M- (Memory Minus):** Subtracts the current value on the display from the stored value in memory.
- **MR (Memory Recall):** Recalls the stored value from memory and displays it on the screen.
- **MC (Memory Clear):** Clears the stored value from memory.

To use the memory functions, first, calculate the value you want to store. Then, press the "M+" key to store the value in memory. You can continue with your calculations, and when you need to recall the stored value, press the "MR" key. To clear the memory, press the "MC" key.

Memory functions can be a real lifesaver when dealing with complex calculations that involve multiple steps or when you need to use the same value repeatedly. By storing intermediate results in memory, you can avoid having to re-enter them, saving time and reducing the risk of errors.

In conclusion, mastering the essential arithmetic operations, including addition, subtraction, multiplication, division, powers, roots, grouping operations, fractions, decimals, percentages, scientific notation, and memory functions, is the foundation for successfully using your scientific calculator. With these fundamental skills under your belt, you'll be well-equipped to tackle more advanced mathematical concepts and solve complex problems in various fields.

In the next chapter, we'll delve into the world of trigonometric functions, exploring how your scientific calculator can help you solve problems involving angles, triangles, and wave phenomena. So, grab your calculator, and let's continue our journey of unlocking its full potential for calculations beyond basic math!

Chapter 3

Trigonometric Functions

Trigonometry, derived from the Greek words "trigonon" (triangle) and "metron" (measure), is a branch of mathematics that deals with the relationships between angles and sides of triangles. In this chapter, we'll explore the fundamental trigonometric functions—sine, cosine, and tangent—and their inverse counterparts. We'll also delve into angle measurements in degrees, radians, and gradians, learn how to solve triangles, understand and apply trigonometric identities, graph trigonometric functions, and uncover their practical applications in various fields.

3.1 Sine, Cosine, and Tangent: Definitions and Applications

The three primary trigonometric functions—sine (sin), cosine (cos), and tangent (tan)—are based on the ratios of sides in a right-angled triangle. Consider a right-angled triangle with an angle θ (theta):

- **Sine (sin θ):** The sine of an angle is the ratio of the length of the side opposite the angle (opposite side) to the length of the hypotenuse (the longest side opposite the right angle).
- **Cosine (cos θ):** The cosine of an angle is the ratio of the length of the side adjacent to the angle (adjacent side) to the length of the hypotenuse.

- **Tangent (tan θ):** The tangent of an angle is the ratio of the length of the opposite side to the length of the adjacent side.

These ratios hold true for any right-angled triangle with the same angle θ, regardless of its size.

Applications of Sine, Cosine, and Tangent:

Trigonometric functions find extensive applications in various fields, including:

- **Physics:** Calculating forces, velocities, and trajectories of objects.
- **Engineering:** Designing structures, analyzing vibrations, and modeling wave phenomena.
- **Surveying:** Measuring distances, angles, and elevations.
- **Navigation:** Determining directions and positions.
- **Astronomy:** Calculating distances to celestial objects and analyzing their movements.

3.2 Inverse Trigonometric Functions: Arcsine, Arccosine, Arctangent

Inverse trigonometric functions, also known as arc functions, are the inverse operations of the sine, cosine, and tangent functions. They are used to find the angle when the ratio of sides is known.

The three inverse trigonometric functions are:

- **Arcsine (arcsin or $\sin^{-1}$):** Returns the angle whose sine is a given value.

- **Arccosine (arccos or cos⁻¹):** Returns the angle whose cosine is a given value.
- **Arctangent (arctan or tan⁻¹):** Returns the angle whose tangent is a given value.

Scientific calculators typically have dedicated keys for accessing inverse trigonometric functions. These functions are essential for solving trigonometric equations, finding angles in triangles, and determining the direction of vectors.

3.3 Degrees, Radians, and Gradians: Converting Between Units

Angles can be measured in three different units: degrees (°), radians (rad), and gradians (grad). Each unit has its own advantages and applications.

- **Degrees:** The most common unit for measuring angles, where a full circle is divided into 360 degrees.
- **Radians:** A unit based on the radius of a circle, where a full circle is equal to 2π radians. Radians are often preferred in calculus and physics as they simplify many trigonometric calculations.
- **Gradians:** A less commonly used unit where a full circle is divided into 400 gradians.

Converting between these units is straightforward:

- **Degrees to Radians:** Multiply the angle in degrees by $\pi/180$.
- **Radians to Degrees:** Multiply the angle in radians by $180/\pi$.

- **Degrees to Gradians:** Multiply the angle in degrees by 10/9.
- **Gradians to Degrees:** Multiply the angle in gradians by 9/10.

Scientific calculators typically have built-in functions for converting between these angle units. Make sure to set your calculator to the correct mode (degrees, radians, or gradians) before performing trigonometric calculations.

3.4 Solving Triangles: Right-Angle and Oblique Triangles

Trigonometry is invaluable for solving triangles — finding unknown side lengths or angles when some information is known.

Right-Angled Triangles:

In a right-angled triangle, we can use trigonometric functions (sine, cosine, tangent) along with the Pythagorean theorem to solve for unknown sides or angles.

The Pythagorean theorem states: $a^2 + b^2 = c^2$, where a and b are the lengths of the legs of the triangle, and c is the length of the hypotenuse.

Oblique Triangles:

For oblique triangles (triangles that do not have a right angle), we use the law of sines and the law of cosines to solve for unknown sides or angles.

- **Law of Sines:** a/sin A = b/sin B = c/sin C, where a, b, and c are the side lengths opposite angles A, B, and C, respectively.
- **Law of Cosines:** $c^2 = a^2 + b^2 - 2ab \cos C$, where a, b, and c are the side lengths, and C is the angle opposite side c.

Scientific calculators simplify the process of solving triangles by allowing you to calculate trigonometric functions, inverse trigonometric functions, and perform complex calculations involving square roots and powers.

3.5 Trigonometric Identities: Understanding and Applying

Trigonometric identities are equations that relate different trigonometric functions to each other. They are powerful tools for simplifying complex trigonometric expressions and solving trigonometric equations.

Some essential trigonometric identities include:

- **Reciprocal Identities:**
    - $\csc \theta = 1/\sin \theta$
    - $\sec \theta = 1/\cos \theta$
    - $\cot \theta = 1/\tan \theta$
- **Quotient Identities:**
    - $\tan \theta = \sin \theta / \cos \theta$
    - $\cot \theta = \cos \theta / \sin \theta$
- **Pythagorean Identities:**
    - $\sin^2 \theta + \cos^2 \theta = 1$
    - $1 + \tan^2 \theta = \sec^2 \theta$
    - $1 + \cot^2 \theta = \csc^2 \theta$

By understanding and applying trigonometric identities, you can manipulate trigonometric expressions, simplify calculations, and solve a wide range of problems in mathematics, physics, engineering, and other fields.

3.6 Graphing Trigonometric Functions

Trigonometric functions can be represented graphically, revealing their periodic nature and characteristic shapes.

The graphs of sine, cosine, and tangent functions exhibit the following features:

- **Period:** The distance between two consecutive peaks or troughs of the graph.
- **Amplitude:** The distance between the midline of the graph and a peak or trough.
- **Phase Shift:** The horizontal shift of the graph from its standard position.

Scientific calculators often have built-in functions for graphing trigonometric functions, allowing you to visualize their behavior and analyze their properties.

Understanding the graphs of trigonometric functions is crucial for understanding their periodic nature, solving trigonometric equations graphically, and analyzing wave phenomena in physics and engineering.

3.7 Practical Applications of Trigonometry: Physics, Engineering

Trigonometry finds widespread applications in various fields, with physics and engineering being prominent examples.

In physics, trigonometry is used to analyze forces, velocities, and accelerations in two or three dimensions. For instance, the components of a force acting at an angle can be calculated using trigonometric functions.

In engineering, trigonometry is essential for designing and analyzing structures, such as bridges, buildings, and machines. It is also used to model vibrations, analyze wave propagation, and design electrical circuits.

The applications of trigonometry extend to various other fields, including:

- **Surveying:** Calculating distances, heights, and angles in land surveying.
- **Navigation:** Determining directions, bearings, and positions for ships, aircraft, and spacecraft.
- **Astronomy:** Calculating distances to celestial objects and predicting their movements.
- **Music:** Analyzing sound waves and synthesizing musical tones.
- **Optics:** Studying the behavior of light and designing lenses and mirrors.

By mastering trigonometry and utilizing the power of your scientific calculator, you open doors to a wide range of exciting applications in various scientific and technical fields.

In conclusion, this comprehensive exploration of trigonometric functions — sine, cosine, tangent, and their inverse counterparts — equips you with a solid foundation in trigonometry. You've learned how to measure angles in different units, solve triangles, manipulate trigonometric expressions using identities, and visualize trigonometric functions through graphing. You've also glimpsed the vast array of practical applications of trigonometry in various fields. As you continue your journey in mathematics, science, or engineering, remember that trigonometry and your scientific calculator are powerful allies that can help you unlock the mysteries of angles, triangles, and wave phenomena.

In the next chapter, we'll delve into the world of logarithmic and exponential functions, exploring how they represent exponential growth and decay, solve equations, and model various real-world phenomena. So, keep your calculator handy and get ready to embark on another exciting chapter of mathematical exploration!

Chapter 4

Logarithmic and Exponential Functions

Logarithmic and exponential functions are intrinsically linked, forming a powerful duo in the world of mathematics. Logarithms offer a unique way to express and manipulate exponents, while exponential functions model growth and decay phenomena. In this chapter, we'll delve into the definitions and properties of logarithms, distinguish between common and natural logarithms, explore antilogarithms, solve logarithmic and exponential equations, and uncover the practical applications of logarithms in various fields. We'll also visualize these functions through graphing and witness their real-world impact in exponential growth and decay models.

4.1 Logarithms: Definitions and Properties

In essence, a logarithm is the inverse operation of exponentiation. It answers the question: "To what power must a base be raised to obtain a given number?"

Formally, the logarithm of a number 'x' to the base 'b' is denoted as $\log_b(x)$ and is defined as:

$\log_b(x) = y$ if and only if $b^y = x$

In this equation:

- 'b' is the base of the logarithm (a positive number not equal to 1)
- 'x' is the argument of the logarithm (a positive number)
- 'y' is the exponent to which the base 'b' must be raised to obtain 'x'

For example, $\log_{10}(100) = 2$, because 10 raised to the power of 2 equals 100.

Properties of Logarithms:

Logarithms possess several important properties that make them valuable tools for mathematical manipulations:

- **Product Rule:** $\log_b(x * y) = \log_b(x) + \log_b(y)$
- **Quotient Rule:** $\log_b(x / y) = \log_b(x) - \log_b(y)$
- **Power Rule:** $\log_b(x^y) = y * \log_b(x)$
- **Change of Base Rule:** $\log_b(x) = \log_a(x) / \log_a(b)$
- **Logarithm of 1:** $\log_b(1) = 0$
- **Logarithm of the Base:** $\log_b(b) = 1$

These properties enable us to simplify logarithmic expressions, solve logarithmic equations, and manipulate exponential relationships.

4.2 Common Logarithms (Base 10) and Natural Logarithms (Base e)

Two specific bases for logarithms are widely used:

- **Common Logarithms (Base 10):** Denoted as $\log_{10}(x)$ or simply $\log(x)$, these logarithms have a base of 10. They are often used in science and engineering to express quantities in orders of magnitude.
- **Natural Logarithms (Base e):** Denoted as $\log_e(x)$ or $\ln(x)$, these logarithms have a base of e, Euler's number (approximately 2.71828). Natural logarithms are prevalent in calculus, physics, and other scientific fields due to their mathematical properties.

Scientific calculators typically have dedicated keys for calculating both common and natural logarithms:

- **log key:** Calculates common logarithms ($\log_{10}$)
- **ln key:** Calculates natural logarithms (ln)

Familiarizing yourself with both types of logarithms is essential for understanding their distinct roles and applications in various disciplines.

4.3 Antilogarithms (Inverse Logarithms)

Antilogarithms, also known as inverse logarithms, are the inverse operations of logarithms. They allow you to find the number whose logarithm is known.

The antilogarithm of a number 'y' to the base 'b' is denoted as $\text{antilog}_b(y)$ and is equivalent to b^y.

For example, $\text{antilog}_{10}(2) = 100$, because 10 raised to the power of 2 equals 100.

Scientific calculators typically have dedicated keys for calculating antilogarithms:

- **10^x key:** Calculates the antilogarithm of common logarithms (10^x)
- **e^x key:** Calculates the antilogarithm of natural logarithms (e^x)

Antilogarithms are essential for solving logarithmic equations and converting logarithmic expressions back into standard numerical form.

4.4 Solving Logarithmic and Exponential Equations

Logarithmic and exponential equations involve expressions with logarithms and exponents. Solving these equations requires understanding the properties of logarithms and utilizing your scientific calculator's functions.

Solving Logarithmic Equations:

To solve a logarithmic equation, the general approach is to isolate the logarithmic term and then apply the appropriate antilogarithm to both sides of the equation.

For example, to solve the equation log(x) = 2, we would apply the antilogarithm with base 10 to both sides, yielding x = 10^2 = 100.

Solving Exponential Equations:

To solve an exponential equation, we can take the logarithm of both sides of the equation to bring down the exponent. We then use the properties of logarithms to isolate the variable.

For instance, to solve the equation 2^x = 8, we take the logarithm of both sides (either common or natural logarithm) and apply the power rule:

log(2^x) = log(8)

x * log(2) = log(8)

x = log(8) / log(2)

x = 3

Your scientific calculator's logarithm and antilogarithm functions make solving logarithmic and exponential equations much easier, allowing you to obtain accurate solutions efficiently.

4.5 Applications of Logarithms: pH, Richter Scale, Decibel Scale

Logarithms find numerous practical applications in various fields, often used to represent quantities that vary over a wide range of values.

pH Scale: The pH scale measures the acidity or alkalinity of a solution. It is a logarithmic scale, where each whole number change in pH represents a tenfold change in the concentration of hydrogen ions (H+). A pH of 7 is neutral, a pH less than 7 is acidic, and a pH greater than 7 is basic (alkaline).

Richter Scale: The Richter scale measures the magnitude of earthquakes. It is a logarithmic scale, where each whole number increase represents a tenfold increase in the amplitude of seismic waves. An earthquake with a magnitude of 6 is ten times stronger than an earthquake with a magnitude of 5.

Decibel Scale: The decibel scale measures the intensity of sound. It is a logarithmic scale, where each 10-decibel increase represents a tenfold increase in sound intensity. A sound of 60 decibels is ten times louder than a sound of 50 decibels.

These are just a few examples of how logarithms are used to quantify and compare values in various scientific and engineering fields. Understanding logarithmic scales is essential for interpreting data and understanding the magnitude of changes represented by these scales.

4.6 Graphing Logarithmic and Exponential Functions

Visualizing logarithmic and exponential functions through graphs provides insights into their behavior and properties.

Logarithmic Functions:

The graph of a logarithmic function is a curve that increases slowly as the input values increase. The curve approaches the y-axis but never touches it, as the logarithm of zero is undefined. The shape of the curve depends on the base of the logarithm.

Exponential Functions:

The graph of an exponential function is a curve that increases rapidly as the input values increase. The curve passes through the point (0, 1), as any number raised to the power of zero equals one. The steepness of the curve depends on the base of the exponent.

Scientific calculators often have built-in functions for graphing logarithmic and exponential functions, allowing you to explore their visual representations and analyze their characteristics.

4.7 Exponential Growth and Decay: Real-World Applications

Exponential functions are powerful tools for modeling phenomena that exhibit exponential growth or decay.

Exponential growth occurs when a quantity increases at a rate proportional to its current value. This type of growth is often observed in populations, investments, and certain chemical reactions.

Exponential decay occurs when a quantity decreases at a rate proportional to its current value. Examples of exponential decay include radioactive decay, the cooling of objects, and the depreciation of assets.

Exponential functions allow us to predict future values, estimate rates of change, and analyze the long-term behavior of these phenomena.

Real-World Applications:

- **Population Growth:** Exponential functions can be used to model the growth of populations, whether it's bacteria in a petri dish or humans in a country.
- **Compound Interest:** The growth of an investment with compound interest can be modeled using an exponential function.
- **Radioactive Decay:** The decay of radioactive isotopes follows an exponential pattern, allowing us to estimate the half-life of isotopes.
- **Drug Elimination:** The elimination of drugs from the body often follows exponential decay, which is

essential for determining dosage and frequency of administration.

By understanding exponential growth and decay, we can make informed decisions, predict future outcomes, and analyze the dynamics of various real-world processes.

In conclusion, this comprehensive exploration of logarithmic and exponential functions has unveiled their significance in the realm of mathematics and their practical applications in diverse fields.

Chapter 5

Statistical Functions

Statistics is the science of collecting, analyzing, interpreting, and presenting data. Scientific calculators are indispensable tools for statisticians and anyone working with data, as they offer a wide range of built-in functions that streamline complex calculations and analysis. In this chapter, we'll explore the essential statistical functions available on your calculator, from measures of central tendency and dispersion to probability calculations, the normal distribution, hypothesis testing, regression analysis, and real-world applications of statistics.

5.1 Mean, Median, and Mode: Measures of Central Tendency

Measures of central tendency provide a single value that represents the "center" or "average" of a dataset. The three most common measures are:

- **Mean (Average):** The mean is calculated by summing all the values in a dataset and dividing by the number of values. It is the most commonly used measure of central tendency, but it can be sensitive to extreme values (outliers).
- **Median:** The median is the middle value in an ordered dataset. If the dataset has an odd number of values, the median is the middle

value. If the dataset has an even number of values, the median is the average of the two middle values. The median is less affected by outliers than the mean.
- **Mode:** The mode is the most frequent value in a dataset. A dataset can have one mode (unimodal), two modes (bimodal), or more than two modes (multimodal).

Scientific calculators typically have built-in functions for calculating the mean, median, and mode. These functions make it easy to analyze the central tendency of your dataset, providing insights into the typical or representative value.

5.2 Range and Standard Deviation: Measures of Dispersion

Measures of dispersion describe the spread or variability of a dataset. Two common measures are:

- **Range:** The range is the difference between the largest and smallest values in a dataset. It is a simple measure of dispersion, but it is sensitive to outliers and doesn't provide information about the distribution of values.
- **Standard Deviation (SD):** The standard deviation is a measure of how spread out the values are from the mean. A small standard deviation indicates that the values are clustered close to the mean, while a large standard deviation indicates that the values are

more spread out. The standard deviation is a widely used measure of dispersion in statistics.

Scientific calculators usually have functions for calculating the range and standard deviation, making it easy to assess the variability of your dataset and understand how much the individual values deviate from the central tendency.

5.3 Probability Calculations: Permutations and Combinations

Probability is the study of random events and their likelihood of occurring. Scientific calculators can assist with probability calculations, particularly those involving permutations and combinations.

- **Permutations:** A permutation is an arrangement of objects in a specific order. The number of permutations of 'n' objects taken 'r' at a time is denoted as nPr and calculated as:

nPr = n! / (n - r)!

- **Combinations:** A combination is a selection of objects without regard to order. The number of combinations of 'n' objects taken 'r' at a time is denoted as nCr and calculated as:

nCr = n! / (r! * (n - r)!)

Scientific calculators often have functions for calculating factorials (!), permutations (nPr), and combinations (nCr). These functions are essential for solving probability problems, analyzing data, and making informed decisions based on the likelihood of events.

5.4 Normal Distribution: Understanding and Applying

The normal distribution, also known as the Gaussian distribution, is a continuous probability distribution that is symmetrical around the mean. It is one of the most important distributions in statistics, as many natural phenomena and human characteristics tend to follow a normal distribution.

The normal distribution is characterized by its mean (μ) and standard deviation (σ). The mean determines the center of the distribution, while the standard deviation determines its spread.

Scientific calculators typically have functions for calculating probabilities associated with the normal distribution. These functions allow you to find the probability of a random variable falling within a certain range of values, given the mean and standard deviation of the distribution.

The normal distribution is widely used in various fields, including:

- **Quality Control:** Assessing the variability of manufactured products.
- **Finance:** Modeling stock prices and other financial instruments.

- **Medicine:** Analyzing patient data and test results.
- **Education:** Evaluating student performance and standardized test scores.

Understanding the normal distribution and its applications is crucial for analyzing data, making predictions, and drawing meaningful conclusions in various fields.

5.5 Hypothesis Testing: Introduction and Basic Concepts

Hypothesis testing is a statistical method for making decisions based on data. It involves formulating a hypothesis (a statement about a population parameter) and then collecting data to test the validity of that hypothesis.

The basic steps of hypothesis testing are:

1. **State the null hypothesis (H0) and alternative hypothesis (Ha).**
2. **Choose a significance level (α).**
3. **Collect data and calculate a test statistic.**
4. **Determine the critical value(s) or p-value.**
5. **Make a decision based on the test statistic and critical value(s) or p-value.**

Scientific calculators can help you perform hypothesis tests by calculating test statistics, critical values, and p-values. However, it's important to understand the underlying concepts of hypothesis testing and choose the appropriate test for your data.

Hypothesis testing is widely used in scientific research, medicine, social sciences, and other fields to evaluate the effectiveness of treatments, interventions, or policies.

5.6 Regression Analysis: Linear and Multiple Regression

Regression analysis is a statistical technique for modeling the relationship between two or more variables. It is used to predict the value of one variable (the dependent variable) based on the values of other variables (the independent variables).

There are two main types of regression analysis:

- **Linear Regression:** Models the relationship between two variables using a straight line.
- **Multiple Regression:** Models the relationship between a dependent variable and two or more independent variables.

Scientific calculators often have functions for performing linear and multiple regression analysis. These functions calculate the regression coefficients, which describe the strength and direction of the relationships between the variables.

Regression analysis is widely used in various fields, including:

- **Economics:** Forecasting economic trends and analyzing the impact of policies.

- **Medicine:** Predicting disease outcomes and evaluating the effectiveness of treatments.
- **Social Sciences:** Studying the relationships between social factors and outcomes.

Understanding regression analysis allows you to model complex relationships between variables, make predictions, and gain insights into the factors that influence outcomes.

5.7 Statistical Applications: Data Analysis and Interpretation

Statistics plays a crucial role in data analysis and interpretation, helping us extract meaningful information from raw data. Scientific calculators are indispensable tools for performing statistical calculations, but their true power lies in the ability to apply statistical methods to real-world problems.

Here are some examples of how statistics and scientific calculators are used in various fields:

- **Market Research:** Analyzing consumer preferences and trends to inform marketing strategies.
- **Clinical Trials:** Evaluating the effectiveness and safety of new drugs and treatments.
- **Environmental Studies:** Assessing the impact of environmental factors on ecosystems and populations.
- **Social Sciences:** Analyzing survey data to understand social attitudes and behaviors.

- **Quality Control:** Monitoring the quality of manufactured products and identifying potential defects.

By applying statistical methods and utilizing the capabilities of your scientific calculator, you can gain valuable insights into data, make informed decisions, and contribute to advancements in various fields.

In conclusion, this comprehensive exploration of statistical functions has equipped you with the knowledge and tools to analyze data, make predictions, and draw meaningful conclusions. Your scientific calculator, with its built-in statistical functions, is an invaluable asset in your statistical toolbox. By mastering these functions and understanding the underlying statistical concepts, you'll be well-prepared to tackle a wide range of challenges in data analysis and interpretation.

Chapter 6

Complex Numbers

Complex numbers extend the number system beyond real numbers, introducing a new dimension with the imaginary unit "i." They play a crucial role in various branches of mathematics, physics, engineering, and other scientific fields. In this chapter, we will explore the concept of complex numbers, their standard form, arithmetic operations, conjugates, modulus, argument, polar form, De Moivre's theorem, and their practical applications in electrical engineering and quantum mechanics.

6.1 Imaginary Unit (i) and Complex Number Form

The imaginary unit, denoted by "i", is defined as the square root of -1:

$i = \sqrt{-1}$

This definition may seem counterintuitive since the square of any real number is always non-negative. However, by introducing the imaginary unit, we open up a new realm of numbers that allow us to solve equations that have no real solutions.

A complex number is a number of the form:

$z = a + bi$

where:

- 'a' is the real part of the complex number
- 'b' is the imaginary part of the complex number

The real part 'a' represents the horizontal component of the complex number on the complex plane, while the imaginary part 'b' represents the vertical component.

Examples of complex numbers:

- 3 + 2i (real part = 3, imaginary part = 2)
- -5 - i (real part = -5, imaginary part = -1)
- 0 + 6i (real part = 0, imaginary part = 6)
- 7 + 0i (real part = 7, imaginary part = 0)

The last example highlights that real numbers are a subset of complex numbers, with the imaginary part being zero.

6.2 Arithmetic Operations with Complex Numbers

Complex numbers can be added, subtracted, multiplied, and divided, much like real numbers. Let's explore these operations:

Addition:

To add two complex numbers, add their real parts and imaginary parts separately:

$(a + bi) + (c + di) = (a + c) + (b + d)i$

Example: $(3 + 2i) + (5 - 4i) = 8 - 2i$

Subtraction:

To subtract two complex numbers, subtract their real parts and imaginary parts separately:

$(a + bi) - (c + di) = (a - c) + (b - d)i$

Example: $(7 - 3i) - (1 + 5i) = 6 - 8i$

Multiplication:

To multiply two complex numbers, use the distributive property and the fact that $i^2 = -1$:

$(a + bi) * (c + di) = (ac - bd) + (ad + bc)i$

Example: $(2 + 3i) * (4 - i) = (8 + 3) + (2 - 12)i = 11 - 10i$

Division:

To divide two complex numbers, multiply both the numerator and denominator by the conjugate of the denominator (which we'll discuss in the next section):

$(a + bi) / (c + di) = [(a + bi) * (c - di)] / [(c + di) * (c - di)]$

Example: $(3 + 4i) / (2 - i) = [(3 + 4i) * (2 + i)] / [(2 - i) * (2 + i)] = (2 + 11i) / 5$

6.3 Conjugate of a Complex Number

The conjugate of a complex number is obtained by changing the sign of the imaginary part. The conjugate of 'a + bi' is denoted as 'a - bi'.

Properties of Conjugates:

- The sum of a complex number and its conjugate is always a real number.
- The product of a complex number and its conjugate is always a real number.
- The conjugate of the sum (or difference) of two complex numbers is the sum (or difference) of their conjugates.
- The conjugate of the product (or quotient) of two complex numbers is the product (or quotient) of their conjugates.

Conjugates are essential for simplifying complex number expressions, especially when dealing with division.

6.4 Modulus and Argument of a Complex Number

The modulus of a complex number, denoted as $|z|$, represents its distance from the origin in the complex plane. It is calculated using the Pythagorean theorem:

$|a + bi| = \sqrt{a^2 + b^2}$

The argument of a complex number, denoted as $\arg(z)$, is the angle between the positive real axis and the line segment connecting the origin to the complex number. It is measured counterclockwise and expressed in radians.

The modulus and argument provide a polar representation of a complex number, which we'll discuss in more detail in the next section.

6.5 Polar Form of Complex Numbers

The polar form of a complex number expresses it in terms of its modulus and argument. It is represented as:

$z = r(\cos\theta + i\sin\theta)$

where:

- 'r' is the modulus of the complex number
- 'θ' is the argument of the complex number

The polar form is particularly useful for multiplying and dividing complex numbers. When multiplying complex numbers in polar form, we multiply their moduli and add their arguments. When dividing complex numbers in polar form, we divide their moduli and subtract their arguments.

6.6 De Moivre's Theorem: Powers and Roots of Complex Numbers

De Moivre's theorem is a powerful tool for calculating powers and roots of complex numbers. It states:

$[r(\cos\theta + i\sin\theta)]^n = r^n(\cos n\theta + i\sin n\theta)$

This theorem allows us to raise a complex number to a power by raising its modulus to that power and multiplying its argument by that power.

De Moivre's theorem is also used to find the nth roots of a complex number. The nth roots of a complex number have the same modulus, but their arguments are equally spaced around a circle.

6.7 Applications of Complex Numbers: Electrical Engineering, Quantum Mechanics

Complex numbers find widespread applications in various fields, including:

- **Electrical Engineering:** Complex numbers are used to represent alternating current (AC) voltages and currents, analyze circuits, and design electrical systems. The imaginary unit "i" is used to represent the phase shift between voltage and current.
- **Quantum Mechanics:** Complex numbers are fundamental to quantum mechanics, which describes the behavior of subatomic particles. The

wave function, which describes the state of a quantum system, is a complex-valued function.

Other Applications:

- **Fluid Dynamics:** Complex numbers are used to model fluid flow and analyze wave phenomena.
- **Control Theory:** Complex numbers are used to design control systems for stabilizing dynamic systems.
- **Signal Processing:** Complex numbers are used to represent and manipulate signals, such as audio and video signals.
- **Fractals:** Complex numbers are used to generate and analyze fractal patterns.

In conclusion, this comprehensive exploration of complex numbers has revealed their fundamental concepts, properties, and diverse applications. By understanding complex numbers and utilizing the functions of your scientific calculator, you gain access to a powerful mathematical tool that opens up new avenues for problem-solving and analysis in a wide range of fields.

Chapter 7

Calculus Functions

Calculus is a branch of mathematics that deals with continuous change. It provides powerful tools for understanding rates of change, accumulating quantities, and solving problems involving motion, optimization, and much more. Scientific calculators are indispensable for performing calculus calculations, enabling us to evaluate derivatives, integrals, and other functions. In this chapter, we'll delve into the fundamental concepts of calculus, explore essential differentiation and integration rules, learn about numerical integration and differential equations, and discover the vast applications of calculus in various fields.

7.1 Derivatives: Instantaneous Rates of Change

The derivative of a function measures its instantaneous rate of change at a specific point. It tells us how fast a function is changing at that particular instant. Geometrically, the derivative represents the slope of the tangent line to the function's graph at a given point.

The derivative of a function $f(x)$ is denoted as $f'(x)$ or dy/dx. The process of finding the derivative is called differentiation.

Applications of Derivatives:

Derivatives have numerous applications in various fields, including:

- **Physics:** Calculating velocities, accelerations, and rates of change of physical quantities.
- **Engineering:** Optimizing designs, analyzing stress and strain, and modeling dynamic systems.
- **Economics:** Analyzing marginal costs, revenues, and profits, and studying economic growth and decay.
- **Finance:** Pricing derivatives, managing risk, and optimizing investment portfolios.
- **Biology:** Modeling population growth, analyzing enzyme kinetics, and studying biological processes.

7.2 Basic Differentiation Rules: Power Rule, Product Rule, Quotient Rule

To efficiently calculate derivatives, we rely on a set of basic differentiation rules:

- **Power Rule:** The derivative of x^n is $nx^{(n-1)}$.
- **Product Rule:** The derivative of the product of two functions $u(x)$ and $v(x)$ is $u'(x)v(x) + u(x)v'(x)$.
- **Quotient Rule:** The derivative of the quotient of two functions $u(x)$ and $v(x)$ is $[u'(x)v(x) - u(x)v'(x)] / [v(x)]^2$.
- **Chain Rule:** The derivative of a composite function $f(g(x))$ is $f'(g(x)) * g'(x)$.
- **Trigonometric Functions:** Derivatives of trigonometric functions follow specific rules, such as the derivative of $\sin(x)$ is $\cos(x)$.

These rules provide a systematic approach for finding derivatives of various functions. However, it's important to practice applying these rules to different types of functions to master their usage.

7.3 Integrals: Area Under a Curve

An integral represents the accumulation of a quantity over an interval. Geometrically, it represents the area under the curve of a function between two points.

The integral of a function f(x) is denoted as ∫f(x) dx. The process of finding the integral is called integration. There are two types of integrals:

- **Indefinite Integrals:** These integrals represent a family of functions that differ by a constant. They are used to find antiderivatives, which are functions whose derivative is the original function.
- **Definite Integrals:** These integrals represent a specific numerical value, which is the area under the curve of the function between two specified points.

Applications of Integrals:

Integrals have a wide range of applications, including:

- **Physics:** Calculating work, energy, and momentum.
- **Engineering:** Determining areas, volumes, and centroids of shapes and objects.
- **Economics:** Analyzing consumer surplus, producer surplus, and economic welfare.

- **Statistics:** Calculating probabilities and expected values.

7.4 Basic Integration Rules: Power Rule, Substitution Rule

Similar to differentiation, integration also has a set of basic rules that facilitate the process of finding integrals:

- **Power Rule:** The integral of $x^n\,dx$ is $(1/(n+1))x^{(n+1)} + C$, where C is the constant of integration.
- **Substitution Rule:** This rule allows us to simplify integrals by substituting a new variable for part of the integrand. It is often used when the integrand contains a function and its derivative.

These rules, along with other techniques like integration by parts and trigonometric substitution, provide a framework for solving a wide variety of integrals. However, not all integrals can be solved analytically using these rules. In such cases, we resort to numerical integration methods, which we'll discuss in the next section.

7.5 Numerical Integration: Approximating Definite Integrals

Numerical integration methods are used to approximate the value of definite integrals when analytical solutions are not feasible. These methods involve dividing the interval of integration into smaller subintervals and approximating the area under the curve in each subinterval using simple geometric shapes, such as rectangles or trapezoids.

Some common numerical integration methods include:

- **Rectangular Rule:** Approximates the area under the curve using rectangles.
- **Trapezoidal Rule:** Approximates the area under the curve using trapezoids.
- **Simpson's Rule:** A more accurate method that approximates the area under the curve using parabolic segments.

Scientific calculators often have built-in functions for numerical integration, allowing you to obtain accurate approximations of definite integrals quickly and easily.

7.6 Differential Equations: Introduction and Basic Concepts

A differential equation is an equation that relates a function to its derivatives. Differential equations are used to model dynamic systems, where the rate of change of a quantity depends on the quantity itself and possibly other variables.

Differential equations can be classified based on their order (the highest order derivative present in the equation) and linearity (whether the equation is linear in the unknown function and its derivatives).

Solving differential equations involves finding a function that satisfies the equation. While analytical solutions are possible for some differential equations, many real-world problems require numerical methods for finding approximate solutions.

Scientific calculators can be helpful for solving simple differential equations and for verifying solutions obtained through other methods.

7.7 Applications of Calculus: Physics, Engineering, Economics

Calculus is a cornerstone of modern science and engineering, with applications spanning a vast range of disciplines. Here are some examples of how calculus is used in different fields:

- **Physics:** Calculus is used to describe motion, analyze forces, calculate work and energy, and model various physical phenomena, such as electromagnetism, thermodynamics, and quantum mechanics.
- **Engineering:** Calculus is essential for designing and analyzing structures, optimizing systems, and modeling dynamic processes in various engineering disciplines, such as mechanical, electrical, civil, and aerospace engineering.
- **Economics:** Calculus is used to analyze economic models, study optimization problems, and predict economic trends. It is also used to model consumer behavior, analyze market dynamics, and evaluate the impact of policies.

Other Applications:

Calculus finds applications in various other fields, including:

- **Biology:** Modeling population growth, analyzing enzyme kinetics, and studying biological processes.
- **Medicine:** Analyzing drug absorption and elimination, modeling tumor growth, and studying the spread of diseases.
- **Computer Science:** Optimizing algorithms, analyzing data, and developing machine learning models.
- **Finance:** Pricing derivatives, managing risk, and optimizing investment portfolios.

In conclusion, this comprehensive exploration of calculus functions and their applications showcases the power and versatility of calculus in solving real-world problems. By understanding the concepts of derivatives, integrals, differential equations, and numerical integration, and utilizing the functions of your scientific calculator, you gain access to a powerful mathematical toolkit that empowers you to analyze change, model dynamic systems, and make informed decisions in various fields.

As you continue your academic or professional journey, remember that calculus is not just a subject to be learned but a language for understanding the world around us. By mastering calculus and utilizing your scientific calculator effectively, you'll be well-prepared to tackle challenges and make significant contributions in your chosen field.

Chapter 8

Advanced Functions and Features

Scientific calculators are not just limited to basic arithmetic and trigonometric functions. They possess a wealth of advanced features that cater to the needs of professionals, scientists, engineers, and students in various fields. In this chapter, we'll delve into these advanced functions, exploring hyperbolic functions, matrix operations, vector calculations, equation solvers, programming capabilities, graphing functions, and the versatile solver functions that simplify financial calculations and unit conversions.

8.1 Hyperbolic Functions: Sinh, Cosh, Tanh

Hyperbolic functions are analogs of trigonometric functions, but instead of being based on a circle, they are based on a hyperbola. The three primary hyperbolic functions are:

- **Hyperbolic sine (sinh x):** Defined as $(e^x - e^{-x}) / 2$.
- **Hyperbolic cosine (cosh x):** Defined as $(e^x + e^{-x}) / 2$.
- **Hyperbolic tangent (tanh x):** Defined as sinh x / cosh x.

These functions find applications in various fields, including:

- **Physics:** Modeling catenary curves (the shape of a hanging chain), analyzing wave propagation, and studying special relativity.
- **Engineering:** Designing transmission lines, analyzing heat transfer, and modeling fluid flow.
- **Mathematics:** Solving differential equations and studying complex analysis.

Most scientific calculators have dedicated keys for calculating hyperbolic functions. If your calculator doesn't have dedicated keys, you can often access these functions through a "hyp" or "2nd" function key.

8.2 Matrix Operations: Addition, Subtraction, Multiplication, Determinants

Matrices are rectangular arrays of numbers, often used to represent systems of equations, transformations, and other mathematical objects. Scientific calculators with matrix capabilities can perform various matrix operations, including:

- **Addition and Subtraction:** Matrices of the same dimensions can be added or subtracted element-wise.
- **Scalar Multiplication:** A matrix can be multiplied by a scalar (a single number) by multiplying each element of the matrix by that scalar.
- **Matrix Multiplication:** Two matrices can be multiplied if the number of columns in the first

matrix equals the number of rows in the second matrix. The resulting matrix has the same number of rows as the first matrix and the same number of columns as the second matrix.
- **Determinant:** The determinant of a square matrix is a number that summarizes important information about the matrix, such as whether it is invertible.

Matrix operations are essential in linear algebra, a branch of mathematics that deals with systems of linear equations and their solutions. They also find applications in various fields, including physics, engineering, computer science, economics, and statistics.

8.3 Vector Calculations: Dot Product, Cross Product

Vectors are quantities that have both magnitude and direction. They are used to represent physical quantities like forces, velocities, and displacements. Scientific calculators with vector capabilities can perform various vector calculations, including:

- **Dot Product:** The dot product of two vectors is a scalar quantity that measures the angle between the vectors and their magnitudes. It is calculated as the sum of the products of the corresponding components of the vectors.
- **Cross Product:** The cross product of two vectors is a vector quantity that is perpendicular to both original vectors. Its magnitude is equal to the area of the parallelogram formed by the two vectors.

Vector calculations are essential in physics, engineering, and other scientific fields that involve the study of forces, motion, and fields. They are also used in computer graphics, animation, and robotics.

8.4 Equation Solvers: Numerical and Symbolic Solutions

Equation solvers are powerful tools that can find solutions to equations, both numerically and symbolically.

- **Numerical Solvers:** These solvers use numerical methods, such as Newton-Raphson iteration, to find approximate solutions to equations. They are particularly useful when analytical solutions are not feasible.
- **Symbolic Solvers:** These solvers use algebraic manipulation to find exact solutions to equations in terms of symbols and variables. They are often used in mathematics, physics, and engineering to derive formulas and solve theoretical problems.

Scientific calculators with equation solvers can be a valuable asset for students, scientists, engineers, and anyone who needs to solve equations regularly. They can save time and effort, especially when dealing with complex equations that would be difficult or impossible to solve manually.

8.5 Programming Functions: Creating Custom Programs

Many scientific calculators offer programming capabilities, allowing users to create custom programs to automate repetitive calculations or perform complex tasks. These programs can be written in a calculator-specific programming language or a standard programming language like BASIC or Python.

Programming functions enable you to extend the functionality of your calculator beyond its built-in features. You can create programs to solve specific problems, perform statistical analysis, generate graphs, or even play games.

8.6 Graphing Functions: 2D and 3D Graphs

Graphing calculators have revolutionized the way we visualize and analyze functions. They allow us to plot functions in two or three dimensions, revealing their behavior and properties.

2D graphing calculators can plot functions of a single variable, such as $y = f(x)$. 3D graphing calculators can plot functions of two variables, such as $z = f(x, y)$.

Graphing calculators offer various features for analyzing graphs, such as finding roots, maxima, minima, and points of intersection. They can also be used to calculate definite integrals and solve differential equations graphically.

Graphing functions are invaluable tools for students, scientists, engineers, and anyone who needs to visualize and analyze mathematical functions. They provide a deeper understanding of functions and their properties, and they can be used to solve problems graphically that would be difficult or impossible to solve analytically.

8.7 Solver Functions: Financial Calculations, Unit Conversions

Many scientific calculators come equipped with solver functions that simplify complex calculations in specific domains, such as finance and unit conversions.

- **Financial Calculators:** These calculators include functions for calculating interest rates, amortization schedules, loan payments, present and future values, and other financial metrics. They are essential tools for financial professionals, investors, and anyone dealing with financial calculations.
- **Unit Conversions:** Scientific calculators often have built-in functions for converting between different units of measurement, such as length, mass, time, temperature, and energy. These functions are invaluable for scientists, engineers, and anyone who needs to work with data in different units.

By leveraging the solver functions of your scientific calculator, you can streamline complex calculations, save time, and reduce the risk of errors. These functions are designed to handle specific types of problems, making them efficient and user-friendly tools for professionals and students alike.

In conclusion, this comprehensive exploration of advanced functions and features has revealed the true potential of your scientific calculator. By mastering these capabilities, you can unlock new levels of efficiency, precision, and problem-solving prowess. Whether you're a student, scientist, engineer, or simply someone who enjoys exploring the world of mathematics, these advanced functions empower you to tackle complex challenges, visualize mathematical concepts, and expand your horizons in the fascinating world of calculations.

As you continue your journey, remember that your scientific calculator is more than just a tool; it's a gateway to a universe of mathematical possibilities. Embrace its power, explore its features, and let it guide you on your path of discovery and innovation.

Chapter 9

Practical Applications in Various Fields

Scientific calculators are not confined to the realm of mathematics classrooms or research labs. Their versatility extends to a wide array of disciplines, empowering professionals and students alike to tackle real-world problems efficiently and accurately. In this chapter, we will explore the diverse applications of scientific calculators in science, engineering, finance, computer science, mathematics, surveying, navigation, and even everyday life. We will uncover how these powerful tools can simplify complex calculations, analyze data, model phenomena, and enhance decision-making in various contexts.

9.1 Science: Chemistry, Physics, Biology

Chemistry:

In chemistry, scientific calculators are indispensable for performing stoichiometric calculations, determining molar masses, balancing chemical equations, calculating concentrations, analyzing reaction kinetics, and evaluating thermodynamic properties. They enable chemists to quickly and accurately compute quantities involved in chemical reactions, ensuring precise measurements and reliable experimental results.

Physics:

Physics heavily relies on mathematical modeling and calculations to describe the behavior of matter and energy. Scientific calculators are used to solve equations of motion, calculate forces, energies, and momenta, analyze wave phenomena, and determine various physical properties like density, pressure, and temperature. They play a crucial role in fields like mechanics, thermodynamics, electromagnetism, optics, and quantum mechanics.

Biology:

While biology may seem less reliant on calculations than physics or chemistry, scientific calculators still find numerous applications in this field. Biologists use calculators to analyze population growth, study enzyme kinetics, determine genetic probabilities, calculate metabolic rates, and assess biodiversity. They also use statistical functions to analyze experimental data and draw meaningful conclusions about biological processes.

9.2 Engineering: Electrical, Mechanical, Civil

Electrical Engineering:

Electrical engineers utilize scientific calculators extensively to analyze circuits, calculate currents, voltages, resistances, and impedances, design electrical systems, and evaluate signal processing algorithms. They leverage the calculator's capabilities to simulate circuits, analyze frequency responses, and optimize the performance of electrical components and systems.

Mechanical Engineering:

Mechanical engineers employ scientific calculators to analyze forces, stresses, and strains in structures, design mechanical systems, calculate thermodynamic properties, and model fluid dynamics. They use calculators to simulate the behavior of machines, optimize designs for efficiency and performance, and analyze the effects of external factors on mechanical systems.

Civil Engineering:

Civil engineers rely on scientific calculators for various calculations related to structural analysis, geotechnical engineering, hydraulics, and transportation engineering. They use calculators to analyze the stability of structures, design foundations, calculate water flow rates, and optimize transportation networks. Calculators are essential tools for ensuring the safety, efficiency, and sustainability of civil engineering projects.

9.3 Finance and Economics: Interest Calculations, Statistics

Finance:

Financial professionals heavily utilize scientific calculators for various financial calculations, such as interest calculations, present and future value analysis, loan amortization schedules, investment returns, and risk assessment. They use calculators to evaluate financial instruments, model market trends, and make informed investment decisions.

Economics:

Economists employ scientific calculators to analyze economic data, model economic behavior, forecast economic trends, and evaluate the impact of policies. They use calculators to calculate economic indicators like GDP, inflation, and unemployment rates. They also use statistical functions to analyze data and draw conclusions about economic relationships and trends.

9.4 Computer Science: Algorithms, Data Analysis

Algorithms:

Computer scientists use scientific calculators to analyze the complexity of algorithms, evaluate their efficiency, and compare different algorithmic approaches. They also use calculators to perform numerical computations involved in algorithm design and analysis.

Data Analysis:

Scientific calculators with statistical functions are essential for analyzing data in computer science. They are used to calculate descriptive statistics, perform hypothesis testing, conduct regression analysis, and evaluate machine learning models. Calculators aid in extracting meaningful insights from data, making informed decisions, and developing effective algorithms and models.

9.5 Mathematics: Algebra, Geometry, Trigonometry

Scientific calculators are fundamental tools for all branches of mathematics. In algebra, they are used to solve equations, simplify expressions, and manipulate polynomials. In geometry, calculators assist with calculating areas, volumes, and angles. In trigonometry, they are used to calculate trigonometric functions, solve triangles, and analyze wave phenomena.

9.6 Surveying and Navigation: Coordinates, Distance Calculations

Surveying:

Surveyors use scientific calculators to perform calculations related to land measurement, mapping, and positioning. They use calculators to determine coordinates, calculate distances, angles, and areas, and analyze survey data. The precision and accuracy of scientific calculators are crucial for ensuring the reliability and accuracy of survey results.

Navigation:

Navigation relies heavily on trigonometry and spherical geometry. Scientific calculators are used by navigators to calculate distances, bearings, and positions, taking into account the curvature of the Earth. They are essential tools for marine navigation, aviation, and space exploration.

9.7 Everyday Life: Budgeting, Conversions, Time Calculations

Scientific calculators are not just for professionals and scientists; they can also be helpful tools for everyday life.

- **Budgeting:** Calculators can assist with managing finances, calculating budgets, tracking expenses, and planning for savings goals.
- **Conversions:** Scientific calculators often have built-in functions for converting between different units of measurement, such as length, mass, volume, temperature, and currency. This can be handy for cooking, traveling, shopping, and other daily activities.
- **Time Calculations:** Calculators can help with time calculations, such as adding or subtracting time intervals, calculating elapsed time, or converting between different time zones.

In conclusion, this comprehensive exploration of the practical applications of scientific calculators in various fields demonstrates their indispensable role in solving real-world problems, analyzing data, modeling phenomena, and making informed decisions. Whether you're a scientist, engineer, financier, computer scientist, mathematician, surveyor, navigator, or simply someone who uses a calculator for everyday tasks, understanding the capabilities of your scientific calculator and its diverse applications can empower you to achieve greater efficiency, accuracy, and success in your endeavors.

Chapter 10

Troubleshooting and Tips

No tool is immune to errors or the occasional hiccup, and scientific calculators are no exception. In this chapter, we'll tackle common calculator issues, guide you through the resetting process, explore resources for finding manuals and online help, discuss firmware upgrades, offer guidance on choosing the right calculator, delve into the world of online calculators and apps, and share valuable tips for effective calculator use. By mastering these troubleshooting techniques and utilizing best practices, you'll ensure that your calculator remains a reliable and efficient companion in your mathematical endeavors.

10.1 Common Calculator Errors: Syntax, Calculation, Overflow

While scientific calculators are powerful tools, they can sometimes produce unexpected results or encounter errors. Understanding common errors and their causes can help you troubleshoot and resolve issues quickly.

Syntax Errors: These occur when you input an expression incorrectly, violating the calculator's syntax rules. Common syntax errors include:

- Mismatched parentheses or brackets

- Incorrect use of functions or operators
- Entering expressions in the wrong order

Calculation Errors: These occur due to incorrect mathematical operations or assumptions. Examples include:

- Dividing by zero
- Taking the square root of a negative number
- Using an incorrect formula or equation

Overflow Errors: These occur when a calculation results in a number that exceeds the calculator's display capacity. In such cases, the calculator may display an error message or a rounded value.

Troubleshooting Tips:

- Double-check your input: Carefully review your expression to ensure that it adheres to the calculator's syntax rules and that all parentheses, brackets, and operators are used correctly.
- Clear the display: If you encounter an error, try clearing the display and re-entering the expression.
- Consult the manual: If you're unsure about the correct syntax or function usage, refer to your calculator's manual for guidance.

10.2 Resetting Your Calculator

Resetting your calculator can often resolve various issues and restore it to its default settings. The process for resetting varies depending on the model and brand of your calculator.

Types of Resets:

- **Soft Reset:** This clears the current calculation and any errors, but it retains stored values and settings.
- **Hard Reset:** This restores the calculator to its factory default settings, erasing all stored values and custom settings.

How to Reset:

Consult your calculator's manual for specific instructions on how to perform a soft or hard reset. In most cases, a combination of key presses is required to initiate the reset process.

Caution:

Before performing a hard reset, back up any important data or settings that you want to retain. A hard reset will erase all stored values and custom settings.

10.3 Finding Manuals and Online Resources

If you need further assistance with your calculator, consult its user manual. The manual provides detailed instructions on how to use the calculator's various functions and features, as well as troubleshooting tips for common errors.

If you've lost your manual, you can often find it online on the manufacturer's website or through a search engine. Many manufacturers also offer online support resources, such as FAQs, tutorials, and forums where you can ask questions and get help from other users.

10.4 Upgrading Your Calculator: Firmware and Software

Some scientific calculators allow you to upgrade their firmware or software to add new features, improve performance, or fix bugs. Firmware updates are typically provided by the manufacturer and can be installed through a computer or directly on the calculator.

Software updates, on the other hand, may involve installing additional software on your computer that interacts with the calculator.

Check the manufacturer's website for information on available firmware and software updates for your specific calculator model.

10.5 Choosing the Right Calculator for Your Needs

With a plethora of scientific calculators available on the market, choosing the right one can be daunting. Consider the following factors when making your decision:

- **Functionality:** Identify the specific functions and features you need for your coursework or profession. Consider whether you need a basic

scientific calculator, a graphing calculator, a programmable calculator, or a specialized calculator for a specific field.
- **Ease of Use:** Choose a calculator with an intuitive interface and easy-to-use keys. Consider the size and layout of the keys, the readability of the display, and the overall design of the calculator.
- **Durability:** If you plan to use your calculator frequently or in demanding environments, opt for a model known for its durability and robustness.
- **Price:** Scientific calculators vary widely in price, from affordable basic models to high-end graphing calculators with advanced features. Set a budget and choose a calculator that offers the best value for your money.
- **Brand Reputation:** Consider the reputation of the brand and its customer support. Choose a brand known for producing reliable calculators with good customer service.

10.6 Utilizing Online Calculators and Apps

In addition to physical calculators, there are numerous online calculators and mobile apps available that can perform various scientific calculations. These online tools often offer a wider range of functions and features than physical calculators, and they can be accessed from any device with an internet connection.

Online calculators and apps can be particularly useful for complex calculations or when you need a specific function that is not available on your physical calculator. However, be sure to choose reputable sources for online calculators and apps to ensure accuracy and security.

10.7 Tips for Effective Calculator Use: Organization, Practice

To maximize the efficiency and accuracy of your calculator use, consider these tips:

- **Organize Your Work:** Before starting a calculation, take a moment to organize your thoughts and plan your steps. Write down the necessary formulas and values to avoid confusion and errors.
- **Use Parentheses and Brackets:** Use parentheses and brackets to clarify the order of operations and avoid ambiguity in complex expressions.
- **Check Your Work:** After completing a calculation, double-check your input and verify the result using mental math or estimation techniques.
- **Practice Regularly:** The more you use your calculator, the more comfortable and proficient you'll become with its functions and features. Practice regularly to improve your speed and accuracy.

By following these tips and applying the troubleshooting techniques discussed in this chapter, you can ensure that your scientific calculator remains a reliable and indispensable tool for all your mathematical endeavors. Remember, the more you practice and explore the capabilities of your calculator, the more confident and proficient you'll become in using it to solve complex problems, analyze data, and unlock the full potential of mathematics.

www.ingramcontent.com/pod-product-compliance
Lightning Source LLC
Chambersburg PA
CBHW070117230526
45472CB00004B/1293